ANIMAL SEASONS

BRIAN WILDSMITH

Oxford University Press

OXFORD NEW YORK TORONTO MELBOURNE

When spring comes, wild flowers bloom in the meadows,
and the new lambs are born.

Young birds hatch out from their eggs,
and their parents feed them in the nest.

Young fawns wobble about
near their mother
as they start to walk.

When summer comes,
the sun grows hotter in the sky.
Harvest mice climb up the corn
and eat it.

Butterflies travel
from flower to flower,
drinking nectar.

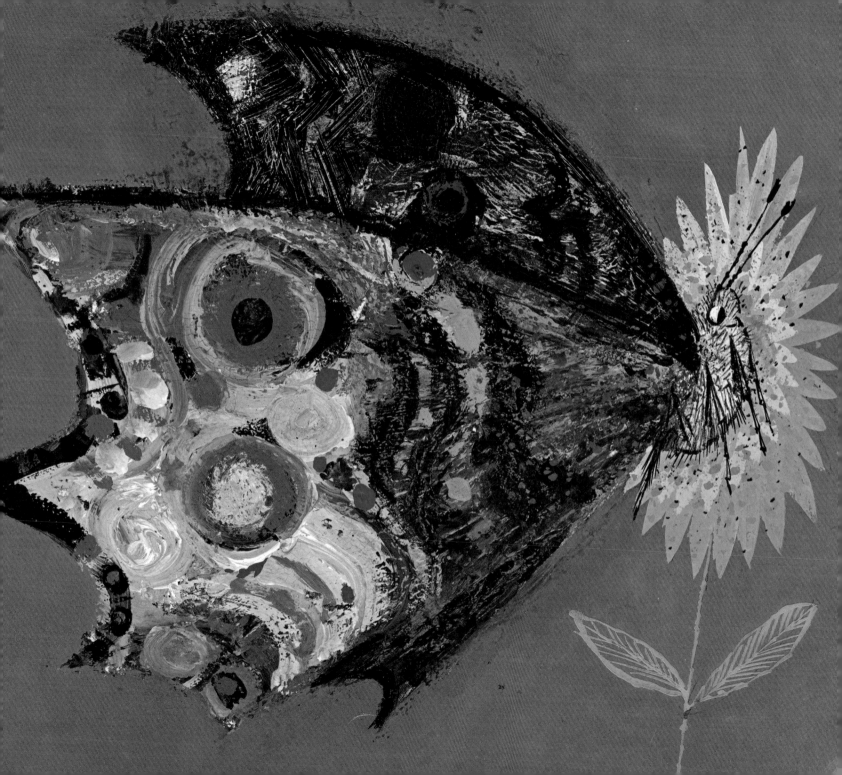

Animals take shelter
in cool places
away from the sun.

When autumn comes,
the leaves turn gold and brown,
and the wind blows them
off the trees.

Squirrels collect nuts and store them for the winter.

Birds gather on roofs and towers,
ready to fly to warmer countries far away.

When winter comes, the rabbits' fur grows long to keep them warm.

Hedgehogs find holes to sleep in all through the winter.

Snow falls and covers everything with white.

Oxford University Press, Walton Street, Oxford OX2 6DP
Oxford is a trade mark of Oxford University Press

© Brian Wildsmith 1980   First published 1980   Reprinted 1983, 1990, 1991
First published in paperback 1991   ISBN 0 19 279730 1 (hardback)   ISBN 0 19 272175 5 (paperback)
Printed in Hong Kong